STRUCTURED LESSONS IN GEOGRAPHY

An Idea Bank For Teachers

Henry T. Conserva

authorHOUSE®

AuthorHouse™
1663 Liberty Drive, Suite 200
Bloomington, IN 47403
www.authorhouse.com
Phone: 1-800-839-8640

First published by AuthorHouse 4/1/2008

ISBN: 978-1-4343-6813-3 (sc)

Library of Congress Control Number: 2008901956

Printed in the United States of America
Bloomington, Indiana

This book is printed on acid-free paper.

In honor of my beloved wife,
Jean F. Dewees.

Preface

I began my career in education in 1951 and ever since my first teaching assignment I have felt the need for good lesson plan ideas. I wanted an "idea bank", or a place to go, such as a bin or a file drawer, where I could pull out lessons that would be appropriate for the class. I didn't have much time in my crowded teaching schedule to sit down and develop lesson plans, so I spent many of my vacation periods searching for lesson plan ideas that I could modify to suit the needs of my classes.

Most of the students I taught in inner city schools needed lessons that were clear and well structured. I found out rather soon in my teaching that discipline problems in class diminished if the assignments required were well understood by the students.

I was not alone in my desire for new ideas for classroom lessons and my colleagues began asking for a copy of my lesson plans to use in their classes. As a result, all of the lesson plans to use in their classes. As a result, all of the lessons presented in this work have been tested in the classroom over a period of years with excellent results. However, these lessons can and should be modified by teachers because every class has a distinctive group personality and lessons usually need to be altered to meet student needs.

Today, social studies includes geography, history, government, anthropology, sociology, and economics as a basic offering of courses of a typical high school studies department. The lessons offered in this "idea bank" are suited to the diversity of courses that teachers are often required to teach even if they have little or no background in one or more of the subdivisions of social studies. I know these structured lessons will be of aid o teachers in meeting the challenge presented by society on both teachers and students to expand academic competence in all fields of study.

Henry Conserva

Table Of Contents

Title: SEARCHING FOR BASINS

Subject Area: physical geography (land forms)

Skills: using an atlas

Completion Time: 20 minutes of class time

Objective: for students to learn the meaning of the term "basin" and be able to locate several basins in an atlas

Comments: this assignment is best suited for an individual student or a pair of students …a class would strain the resources available to most teachers

References: atlases

Extension Activities: students could list other basins not included in this lesson

Notes:

Searching For Basins

Directions: Place an (x) in the box that connects each basin with the nation where it is found

Definition - basin: the land drained by a river; any depression in the earth's surface

Nations Where Basins Are Found

Basins	Brazil	U.S.A. (Nevada)	west central France	northwestern China	Russia	Australia	U.S.A. (Wyoming)	south central China	southwestern France	Democratic Republic of the Congo
1. Great Basin		X								
2. Amazon Basin	X									
3. Pechora Basin					X					
4. Congo Basin										X
5. Dzungarian Basin				X						
6. Great Artesian Basin						X				
7. Wyoming Basin							X			
8. Paris Basin			X							
9. Sichuan Basin								X		
10. Aquitaine Basin									X	

SEARCHING FOR BASINS

Directions: Place an (x) in the box that connects each basin with the nation where it is found

Definition - basin: the land drained by a river; any depression in the earth's surface

NATIONS WHERE BASINS ARE FOUND

BASINS	Brazil	U.S.A. (Nevada)	west central France	northwestern China	Russia	Australia	U.S.A. (Wyoming)	south central China	southwestern France	Democratic Republic of the Congo
1. Great Basin										
2. Amazon Basin										
3. Pechora Basin										
4. Congo Basin										
5. Dzungarian Basin										
6. Great Artesian Basin										
7. Wyoming Basin										
8. Paris Basin										
9. Sichuan Basin										
10. Aquitaine Basin										

Title: SEARCHING FOR CAPES

Subject Area: physical geography (land forms)

Skills: using an atlas

Completion Time: 20 minutes of class time

Objective: for students to learn the meaning of the term "cape" and to be able to locate several of them in an atlas

Comments: this assignment is best suited for an individual student or a pair of students… larger groups often strain the resources of many teachers

References: atlases

Extension Activities: students could list other capes not included in this lesson

Notes:

SEARCHING FOR CAPES

Directions: Place an (x) in the box that connects each cape to the nation where it is found

Definition – cape: a pointed piece of land jutting out into the sea

NATIONS WHERE CAPES ARE FOUND

CAPES	New Zealand	Senegal	Australia	U.S.A.	Canada	Somalia	South Africa	Mexico	Chile	Portugal
1. North Cape	X									
2. Cape Leevwin			X							
3. Cape Guardafui						X				
4. Cape of Good Hope							X			
5. Cape Verde		X								
6. Cape St. Vincent										X
7. Cape Horn									X	
8. Cape San Lucas								X		
9. Cape Race					X					
10. Cape Cod				X						

SEARCHING FOR CAPES

Directions: Place an (x) in the box that connects each cape to the nation
where it is found

Definition – cape: a pointed piece of land jutting out into the sea

NATIONS WHERE CAPES ARE FOUND

CAPES	New Zealand	Senegal	Australia	U.S.A.	Canada	Somalia	South Africa	Mexico	Chile	Portugal
1. North Cape										
2. Cape Leevwin										
3. Cape Guardafui										
4. Cape of Good Hope										
5. Cape Verde										
6. Cape St. Vincent										
7. Cape Horn										
8. Cape San Lucas										
9. Cape Race										
10. Cape Cod										

Title: SEARCHING FOR DESERTS

Subject Area: physical geography (land forms)

Skills: using an atlas

Completion Time: 20 minutes of class time

Objective: for students to learn the meaning of the term "desert" and to be able to locate several of them in an atlas

Comments: this assignment is best suited for individual students or students working as a pair...a whole class might strain available resources

References: atlases

Extension Activities: students could list other deserts not included in this lesson

Notes:

Searching For Deserts

Directions: Place an (x) in the box that connects each nation or nation set with the desert found there

Definition – desert: an almost barren tract of land in which precipitation is so small that it will not adequately support vegetation

Nations Where Deserts Are Found

Deserts	Turkmenistan	China	Mongolia	Algeria/Mauretania	India	Sudan	Chile	Botswana	U.S.A.	Australia
1. Great Sandy Desert										X
2. Tala Makan Desert		X								
3. Gobi Desert			X							
4. Kara Kum Desert	X									
5. Nubian Desert						X				
6. Kalahari Desert								X		
7. Iguidi Desert				X						
8. Atacama Desert							X			
9. Mojave Desert									X	
10. Great Indian Desert					X					

SEARCHING FOR DESERTS

Directions: Place an (x) in the box that connects each nation or nation set with the desert found there

Definition – desert: an almost barren tract of land in which precipitation is so small that it will not adequately support vegetation

NATIONS WHERE DESERTS ARE FOUND

DESERTS	Turkmenistan	China	Mongolia	Algeria/Mauretania	India	Sudan	Chile	Botswana	U.S.A.	Australia
1. Great Sandy Desert										
2. Tala Makan Desert										
3. Gobi Desert										
4. Kara Kum Desert										
5. Nubian Desert										
6. Kalahari Desert										
7. Iguidi Desert										
8. Atacama Desert										
9. Mojave Desert										
10. Great Indian Desert										

Title: SEARCHING FOR ISLANDS

Subject Area: physical geography (land forms)

Skills: using an atlas

Completion Time: 20 minutes of class time

Objective: for students to learn the meaning of the term "island" and to be able to locate several of them in an atlas

Comments: this assignment is best suited for an individual student or a pair of students… a whole class might strain the available resources of the library

References: atlases

Extension Activities: students could list other islands not included in this lesson

Notes:

SEARCHING FOR ISLANDS

Directions: Place an (x) in the box that connects each island to its national
location

Definition – island: a land mass, especially one smaller than a continent
entirely surrounded by water

NATIONS WHERE ISLANDS ARE FOUND

ISLANDS	Russia	Chile	China	Australia	Cuba	Tanzania	Greece	Indonesia	Canada	Yemen Aden
1. Chiloe Island		X								
2. Isle of Youth					X					
3. Mansel Island									X	
4. Evvoia Island							X			
5. Pemba Island						X				
6. Socotra Island										X
7. Hainan Island			X							
8. Wrangel Island	X									
9. Sumba Island								X		
10. Groote Eylandt				X						

SEARCHING FOR ISLANDS

Directions: Place an (x) in the box that connects each island to its national location

Definition – island: a land mass, especially one smaller than a continent entirely surrounded by water

NATIONS WHERE ISLANDS ARE FOUND

ISLANDS	Russia	Chile	China	Australia	Cuba	Tanzania	Greece	Indonesia	Canada	Yemen Aden
1. Chiloe Island										
2. Isle of Youth										
3. Mansel Island										
4. Evvoia Island										
5. Pemba Island										
6. Socotra Island										
7. Hainan Island										
8. Wrangel Island										
9. Sumba Island										
10. Groote Eylandt										

Title: SEARCHING FOR MOUNTAINS

Subject Area: physical geography (land forms)

Skills: using an atlas

Completion Time: 20 minutes of class time

Objective: for students to learn the meaning of the term "mountain" and to be able to locate several of them in an atlas

Comments: this assignment is best suited for the individual student or for a pair of students… classroom resources may not allow for more

References: atlases

Extension Activities: students could add to the list of mountains several not included in this less

Notes:

SEARCHING FOR MOUNTAINS

Directions: Place an (x) in that box that connects each mountain with the nation where it is found

Definition – mountain: a mass of land considerably higher than its surroundings

NATIONS WHERE MOUNTAINS ARE FOUND

MOUNTAINS	France	Australia	U.S.A.	New Zealand	Indonesia	Ethiopia	Georgia	Canada	Mexico	Ecuador
1. Mt. Elbert			X							
2. Mt. Robson								X		
3. Orizaba									X	
4. Mt. Chimborazo										X
5. Mt. Blanc	X									
6. Mt. Elbrus							X			
7. Ras Dashen						X				
8. Jaya Pech					X					
9. Mt. Kosciusko		X								
10. Mt. Cook				X						

Searching For Mountains

Directions: Place an (x) in the box that connects each mountain with the nation where it is found

Definition – mountain: a mass of land considerably higher than its surroundings

<u>Nations Where Mountains Are Found</u>

Mountains	France	Australia	U.S.A.	New Zealand	Indonesia	Ethiopia	Georgia	Canada	Mexico	Ecuador
1. Mt. Elbert										
2. Mt. Robson										
3. Orizaba										
4. Mt. Chimborazo										
5. Mt. Blanc										
6. Mt. Elbrus										
7. Ras Dashen										
8. Jaya Pech										
9. Mt. Kosciusko										
10. Mt. Cook										

Title: SEARCHING FOR MOUNTAIN RANGES

Subject Area: physical geography (land forms)

Skills: using an atlas

Completion Time: 20 minutes of class time

Objective: for students to learn the meaning of the term "mountain range" and be able to locate several of them in an atlas

Comments: this lesson is best given to individual students or a pair so that atlas resources won't be strained

References: atlases

Extension Activities: students could list mountain ranges not included in this lesson

Notes:

Searching For Mountain Ranges

Directions: Place an (x) in the box that connects each mountain range with the nation where it is found

Definition – mountain range: a chain of mountains

Nations Where Mountain Ranges Are Found

Mountain Ranges	Greece	Russia	Canada	China	Brazil	United Kingdom	U.S.A.	Australia	New Zealand	South Africa
1. Southern Alps									X	
2. Hamersley Ranges								X		
3. Daxue Mountains				X						
4. Baikal Mountains		X								
5. Draksenburg Mts.										X
6. Grampian Mountains						X				
7. Pindus Mountains	X									
8. Sierra Espinhaco					X					
9. Mackensie Mountains			X							
10. Sacramento Mts.							X			

Searching For Mountain Ranges

Directions: Place an (x) in the box that connects each mountain range with the nation where it is found

Definition – mountain range: a chain of mountains

Mountain Ranges	Greece	Russia	Canada	China	Brazil	United Kingdom	U.S.A.	Australia	New Zealand	South Africa
1. Southern Alps										
2. Hamersley Ranges										
3. Daxue Mountains										
4. Baikal Mountains										
5. Draksenburg Mts.										
6. Grampian Mountains										
7. Pindus Mountains										
8. Sierra Espinhaco										
9. Mackensie Mountains										
10. Sacramento Mts.										

Title: SEARCHING FOR PENINSLAS

Subject Area: physical geography (land forms)

Skills: using an atlas

Completion Time: 20 minutes of class time

Objective: for students to understand the meaning of the term "peninsula" and be able to locate several of them in an atlas

Comments: this lesson is best suited for individual student work or a pair of students... too many students might strain available resources

References: atlases

Extension Activities: students could list other peninsulas other than those included in this lesson

Notes:

Searching For Peninsulas

Directions: Place an (x) in the box that connects each peninsula to the place where it is found

Definition – peninsula: a stretch of land almost surrounded by water

Places Where Peninsulas Are Found

Peninsulas	by the Red Sea	between the Caribbean Sea & the Gulf of Mexicao	Between the Black Sea & the Sea of Azov	northern Australia	by the Strait of Malacca	by the Gulf of San Matias	between the Gulf of Mexico & the Atlantic Ocean	by the Bering Sea	by the Gulf of Aden	by Hudson Bay
1. Cape York Peninsula				X						
2. Crimea Peninsula			X							
3. Alaska Peninsula								X		
4. Ungava Peninsula										X
5. Florida Peninsula							X			
6. Yucatan Peninsula		X								
7. Valdes Peninsula						X				
8. Somali Peninsula									X	
9. Sinai Peninsula	X									
10. Malay Peninsula					X					

Searching For Peninsulas

Directions: Place an (x) in the box that connects each peninsula to the place where it is found

Definition – peninsula: a stretch of land almost surrounded by water

Places Where Peninsulas Are Found

Peninsulas	by the Red Sea	between the Caribbean Sea & the Gulf of Mexicao	Between the Black Sea & the Sea of Azov	northern Australia	by the Strait of Malacca	by the Gulf of San Matias	between the Gulf of Mexico & the Atlantic Ocean	by the Bering Sea	by the Gulf of Aden	by Hudson Bay
1. Cape York Peninsula										
2. Crimea Peninsula										
3. Alaska Peninsula										
4. Ungava Peninsula										
5. Florida Peninsula										
6. Yucatan Peninsula										
7. Valdes Peninsula										
8. Somali Peninsula										
9. Sinai Peninsula										
10. Malay Peninsula										

Title: SEARCHING FOR PLAINS

Subject Area: physical geography (land forms)

Skills: using an atlas

Completion Time: 20 minutes of class time

Objective: for students to understand the meaning of the term "plains" and to be able to locate several of them in an atlas

Comments: this lesson is best suited for an individual or pair of students...too many students may stress the resources available to the teacher

References: atlases

Extension Activities: students could list other plains not included in this lesson

Notes:

Searching For Plains

Directions: Place an (x) in the box that connects each plain with the place where it is found

Definition – plain: an extensive, level, treeless land region

Nations Where Plains Are Found

Plains	China	Argentina	India	southern U.S.A.	Brazil	eastern Russia	Colombia\Venezuela	Canada/U.S.A	western USSR	Australia
1. Manchurian Plain	X									
2. Ganges Plain			X							
3. West Siberian Plain						X				
4. Nullarbor Plain										X
5. Oka-Don Plain									X	
6. Pampas		X								
7. Great Plains								X		
8. Coastal Plain				X						
9. Llanos							X			
10. Campos					X					

SEARCHING FOR PLAINS

Directions: Place an (x) in the box that connects each plain with the place
where it is found

Definition – plain: an extensive, level, treeless land region

NATIONS WHERE PLAINS ARE FOUND

PLAINS	China	Argentina	India	southern U.S.A.	Brazil	eastern Russia	Colombia\Venezuela	Canada/U.S.A	western USSR	Australia
1. Manchurian Plain										
2. Ganges Plain										
3. West Siberian Plain										
4. Nullarbor Plain										
5. Oka-Don Plain										
6. Pampas										
7. Great Plains										
8. Coastal Plain										
9. Llanos										
10. Campos										

Title: SEARCHING FOR PLATEAUS

Subject Area: physical geography (land Forms)

Skills: using an atlas

Objective: for students to understand the meaning of the term "plateau" and to be able to locate several of them in an an atlas

Comments: this lesson is best suited for an individual student of a pair of students…too many students might strain the available resources

References: atlases

Extension Activities: students could list other plateaus not included in this lesson

Notes:

SEARCHING FOR PLATEAUS

Directions: Place an (x) in the box that connects each nation with the plateau found there

Definition – plateau: an extensive, level or mainly level area of elevated land

NATIONS WHERE PLATEAUS ARE FOUND

PLATEAUS	Democratic Republic of the Congo	Northwaestern U.S.A.	Mexico	Nigeria	Australia	Russia	India	Brazil	Eastern U.S.A.	China
1. Kimberly Plateau					X					
2. Plateau of Tibet										X
3. Central Siberian Plateau						X				
4. Katanga Plateau	X									
5. Jos Plateau				X						
6. Mato Grasso Plateau								X		
7. Plateau of Mexico			X							
8. Allegheny Plateau									X	
9. Deccan Plateau							X			
10. Colombia Plateau		X								

Searching For Plateaus

Directions: Place an (x) in the box that connects each nation with the plateau found there

Definition – plateau: an extensive, level or mainly level area of elevated land

Nations Where Plateaus Are Found

Plateaus	Democratic Republic of the Congo	Northwaestern U.S.A.	Mexico	Nigeria	Australia	Russia	India	Brazil	Eastern U.S.A.	China
1. Kimberly Plateau										
2. Plateau of Tibet										
3. Central Siberian Plateau										
4. Katanga Plateau										
5. Jos Plateau										
6. Mato Grasso Plateau										
7. Plateau of Mexico										
8. Allegheny Plateau										
9. Deccan Plateau										
10. Colombia Plateau										

Title: SEARCHING FOR BAYS

Subject Area: physical geography (water forms)

Skills: using an atlas

Completion Time: 20 minutes of class time

Objective: for students to understand the meaning of the term "bay" and to be able to locate several of them in an atlas

Comments: this assignment is best suited for an individual student or a pair of students if there are a few atlases available

References: atlases

Extension Activities: students could list bays not included in this lesson

Notes:

Searching For Bays

Directions: Place an (x) in the box that connects each bay to the nation
where it is found

Definition – bay: a wide indentation into the land, formed by the sea or a
lake

Nations Where Bays Are Found

Bays	France/Spain	Namibia	western U.S.A.	eastern Canada	U.S.A./Canada	New Zealand	Australia	northwestern Canada	India/Myanmar (Burma)	Argentina
1. Monterey Bay			X							
2. Bay of Plenty						X				
3. Roebuck Bay							X			
4. Bay of Bengal									X	
5. Bay of Biscay	X									
6. Walvis Bay		X								
7. Bianca Bay										X
8. Bay of Fundy					X					
9. Mackenzie Bay								X		
10. Hudson Bay				X						

Searching For Bays

Directions: Place an (x) in the box that connects each bay to the nation where it is found

Definition – bay: a wide indentation into the land, formed by the sea or a lake

Nations Where Bays Are Found

Bays	France/Spain	Nambia	western U.S.A.	eastern Canada	U.S.A.\Canada	New Zealand	Australia	northwestern Canada	India/Myanmar (Burma)	Argentina
1. Monterey Bay										
2. Bay of Plenty										
3. Roebuck Bay										
4. Bay of Bengal										
5. Bay of Biscay										
6. Walvis Bay										
7. Bianca Bay										
8. Bay of Fundy										
9. Mackenzie Bay										
10. Hudson Bay										

Title: SEARCHING FOR GULFS

Subject Area: physical geography (water forms)

Skills: using an atlas

Completion Time: 20 minutes of class time

Objective: for students to understand the meaning of the term "gulf" and to be able to locate several of them in an atlas

Comments: this lesson is best suited for an individual student or a pair of students… classroom resources may be too limited for a whole class to do this at once

References: atlases

Extension Activities: students could list gulfs not included in this lesson

Notes:

Searching For Gulfs

Directions: Place an (x) in the box that connects each gulf with the nation or region where it's found

Definition – gulf: a large deep bay; an extensive inlet penetrating far into the land

Nations Or Regions In Which Gulfs Are Found

Gulfs	China	Australia	Argentina	Sweden/Finland	Thailand	West Africa	Ecuador	Mexico	Canada	France
1. Joseph Bonaparte Gulf		X								
2. Gulf of Thailand					X					
3. Bo Gulf	X									
4. Gulf of Guinea						X				
5. Gulf of Lion										X
6. Gulf of Bothnia				X						
7. Gulf of San Jorge			X							
8. Gulf of Guayaquil							X			
9. Gulf of Tehuantepec								X		
10. Queen Maud Gulf									X	

SEARCHING FOR GULFS

Directions: Place an (x) in the box that connects each gulf with the nation or region where it's found

Definition – gulf: a large deep bay; an extensive inlet penetrating far into the land

NATIONS OR REGIONS IN WHICH GULFS ARE FOUND

GULFS	China	Australia	Argentina	Sweden/Finland	Thailand	West Africa	Ecuador	Mexico	Canada	France
1. Joseph Bonaparte Gulf										
2. Gulf of Thailand										
3. Bo Gulf										
4. Gulf of Guinea										
5. Gulf of Lion										
6. Gulf of Bothnia										
7. Gulf of San Jorge										
8. Gulf of Guayaquil										
9. Gulf of Tehuantepec										
10. Queen Maud Gulf										

Title: SEARCHING FOR LAKES

Subject Area: physical geography (water forms)

Skills: using an atlas

Completion Time: 20 minutes of class time

Objective: for students to learn the meaning of the term "lake" and to be able to locate several of them in an atlas

Comments: this lesson is best suited to individual student assignments or students working in pairs where atlas resources are limited

References: atlases

Extension Activities: students could list other lakes not included in this lesson

Notes:

Searching For Lakes

Directions: Place an (x) in the box that connects each lake with the nation
where it is found

Definition – lake: a large inland body of fresh or salt water

Nations Where Lakes Are Found

Lakes	Russia/Estonia	Canada	Nicaragua	Venezuela	Sweden	Zambia	Kampuchea (Cambodia)	China	Australia	Mexico
1. Lake Peipus	X									
2. Dubwat Lake		X								
3. Lake Managua			X							
4. Lake Dissapointment									X	
5. Lake Maracaibo				X						
6. Lake Malaren					X					
7. Tonle Sap							X			
8. Poyang Lake								X		
9. Lake Mweru						X				
10. Lake Chapala										X

Searching For Lakes

Directions: Place an (x) in the box that connects each lake with the nation where it is found

Definition – lake: a large inland body of fresh or salt water

Lakes	Russia/Estonia	Canada	Nicaragua	Venezuela	Sweden	Zambia	Kampuchea (Cambodia)	China	Australia	Mexico
1. Lake Peipus										
2. Dubwat Lake										
3. Lake Managua										
4. Lake Dissapointment										
5. Lake Maracaibo										
6. Lake Malaren										
7. Tonle Sap										
8. Poyang Lake										
9. Lake Mweru										
10. Lake Chapala										

Title: SEARCHING FOR RIVERS

Subject Area: physical geography (water forms)

Skills: using an atlas

Completion Time: 20 minutes of class time

Objective: for students to learn the meaning of the term "river" and to be able to locate several of them in an atlas

Comments: this lesson is best suited to assignments to an individual student or a pair of students…class room resources may be strained if all students need an atlas

References: atlases

Extension Activities: students could make a list of other rivers that are not included in this lesson

Notes:

SEARCHING FOR RIVERS

Directions: Place an (x) in the box that connects each river to its national location

Definition – river: a large natural stream of water emptying into an ocean, lake or other body of water

NATIONS WHERE RIVERS ARE FOUND

RIVERS	Romania	Russia	Spain/Portugal	Nicaragua	Chad	Bolivia	Canada	Australia	Angola	Mexico
1. Pely River								X		
2. Cuanza River									X	
3. Chari River					X					
4. Guadiana River			X							
5. Mures River	X									
6. Yenisey River		X								
7. Grande River				X						
8. Mamore River						X				
9. Lerma River										X
10. Feuilles River							X			

38

SEARCHING FOR RIVERS

Directions: Place an (x) in the box that connects each river to its national
location

Definition – river: a large natural stream of water emptying into an ocean,
lake or other body of water

NATIONS WHERE RIVERS ARE FOUND

RIVERS	Romania	Russia	Spain/Portugal	Nicaragua	Chad	Bolivia	Canada	Australia	Angola	Mexico
1. Pely River										
2. Cuanza River										
3. Chari River										
4. Guadiana River										
5. Mures River										
6. Yenisey River										
7. Grande River										
8. Mamore River										
9. Lerma River										
10. Feuilles River										

Title: SEARCHING FOR SEAS

Subject Area: physical geography (water forms)

Skills: using an atlas

Completion Time: 20 minutes of class time

Objective: for students to understand the meaning of the term "sea" and to be able to locate several of them in an atlas

Comments: this lesson is best suited to be done by an individual student or a pair of students…greater numbers of students may strain the resources of some classrooms

References: atlases

Extension Activities: students could list other seas not included in this lesson

Notes:

Searching For Seas

<u>Directions:</u> Place an (x) in the box that connects each sea with the nation by which it is found

<u>Definition -</u> One of the smaller divisions of the oceans; a large expanse of inland salt water

Nations By Which Seas Are Found

SEAS	Italy	Australia/New Zealand	Indonesia/Australia	China/Korea	United Kingdom/Ireland	Canada	Russia	Spain	India	Indonesia
1. Arafura Sea			X							
2. Java Sea										X
3. Tasman Sea		X								
4. Sea of Okhotsk							X			
5. Laccadive Sea									X	
6. Ligurian Sea	X									
7. Belearic Sea								X		
8. Irish Sea					X					
9. Labrador Sea						X				
10. Yellow Sea				X						

Searching For Seas

Directions: Place an (x) in the box that connects each sea with the nation by which it is found

Definition - One of the smaller divisions of the oceans; a large expanse of inland salt water

Seas	Italy	Australia/New Zealand	Indonesia/Australia	China/Korea	United Kingdom/Ireland	Canada	Russia	Spain	India	Indonesia
1. Arafura Sea										
2. Java Sea										
3. Tasman Sea										
4. Sea of Okhotsk										
5. Laccadive Sea										
6. Ligurian Sea										
7. Belearic Sea										
8. Irish Sea										
9. Labrador Sea										
10. Yellow Sea										

Title: SEARCHING FOR STRAITS

Subject Area: physical geography (water forms)

Skills: using an atlas

Completion Time: 20 minutes of class time

Objective: for students to understand the meaning of the term "straight" and to be able to locate several of them in an atlas

Comments: this lesson is best suited to an individual student or a pair of students…too many students may strain classroom resources

References: atlases

Extension Activities: students could list other straits not included in this lesson

Notes:

Seaching For Straits

Directions: Place an (x) in the box that connects each strait to the place where it is found

Definition – strait: a narrow stretch of sea connecting two extensive areas of sea

Nations Where Straits Are Found

STRAITS	Canada	Indonesia	Taiwan/Philippines	United Kingdom/France	Argentina/Chile	Spain/Morocco	Taiwan/China	Australia	Korea/Japan	New Zealand
1. Cook Strait										X
2. Bass Strait								X		
3. Karimaia Strait		X								
4. Korea Strait									X	
5. Formosa Strait							X			
6. Luzon Strait			X							
7. Strait of Gibraltar						X				
8. Strait of Dover				X						
9. Strait of Magellan					X					
10. Hudson Strait	X									

SEACHING FOR STRAITS

Directions: Place an (x) in the box that connects each strait to the place where it is found

Definition – strait: a narrow stretch ofsea connecting two extensive areas of sea

NATIONS WHERE STRAITS ARE FOUND

STRAITS	Canada	Indonesia	Taiwan/Philippines	United Kingdom/France	Argentina/Chile	Spain/Morocco	Taiwan/China	Australia	Korea/Japan	New Zealand
1. Cook Strait										
2. Bass Strait										
3. Karimaia Strait										
4. Korea Strait										
5. Formosa Strait										
6. Luzon Strait										
7. Strait of Gibraltar										
8. Strait of Dover										
9. Strait of Magellan										
10. Hudson Strait										

Title: NATIONS AND THEIR HEMISPHERES

Subject Area: physical geography (divisions of the earth's surface)

Skills: using the maps and atlases

Completion Time: 20 minutes of class time

Comments: maps of the world and atlases can be used by the students to do this lesson but globes would be most effective at least for an individual student or a pair of students

References: maps, globes, atlases

Extension Activities: students could make a list of nations other than those in this lesson and locate the nations in t heir correct hemispheres

Notes:

NATIONS AND THEIR HEMISPHERES

Directions: Place an (x) in the boxes that connect each nation or region with its set of hemispheric locations

Definition – hemisphere: the half of the earth's surface, formed when a plane through its center bisects the earth

HEMISPHERIC SECTIONS

NATIONS	Northern Hemisphere	Southern Hemisphere	Eastern Hemisphere	Western Hemisphere						
1. Canada	X			X						
2. Botswana		X	X							
3. Australia		X	X							
4. Brazil		X		X						
5. China	X		X							
6. Tanzania		X	X							
7. Sweden	X		X							
8. Mexico	X			X						
9. India	X		X							
10. Chad	X		X							

Nations And Their Hemispheres

Directions: Place an (x) in the boxes that connect each nation or region with its set of hemispheric locations

Definition – hemisphere: the half of the earth's surface, formed when a plane through its center bisects the earth

Hemispheric Sections

Nations	Northern Hemisphere	Southern Hemisphere	Eastern Hemisphere	Western Hemisphere						
1. Canada										
2. Botswana										
3. Australia										
4. Brazil										
5. China										
6. Tanzania										
7. Sweden										
8. Mexico										
9. India										
10. Chad										

Title: LOCATING NATIONS USING COORDINATES

Subject Area: physical geography (locating nations on the earth's surface)

Skills: using maps, globes and atlases

Completion Time: a 45 minute period of class time

Objective: for students to be able to locate places on a map or globe showing the earth's surface by use of coordinates

Comments: globes may be better than either maps or atlases for this lesson

References: maps, globes and atlases

Extension Activities: students could list nations not included in these four exercises and assign the nations a set of coordinates appropriate to the locations of the nations

Notes:

Exercise #1
Locating Nations Using Coordinants

Directions: Place an (x) in the box that connects each nation with its coordinants

Definition – coordinants: one of a set of numbers that determines the location of a point in the space of a given dimension

Nations

Coordinates Lat.	Long.	Panama	Peru	Uruguay	Italy	Finland	Democratic Republic of the Congo	South Africa	Chad	Thailand	Taiwan
1. 23½ N	120 E										X
2. 15 S	75 W		X								
3. 65 N	25 E					X					
4. 8 N	82 W	X									
5. 20 N	100 E									X	
6. 10 N	20 E								X		
7. 35 S	55 W			X							
8. 0	20 E						X				
9. 45 N	10 E				X						
10. 30 S	30 E							X			

Exercise #1
Locating Nations Using Coordinants

Directions: Place an (x) in the box that connects each nation with its coordinants

Definition – coordinants: one of a set of numbers that determines the location of a point in the space of a given dimension

NATIONS

COORDINATES Lat.	Long.	Panama	Peru	Uruguay	Italy	Finland	Democratic Republic of the Congo	South Africa	Chad	Thailand	Taiwan
1. 23½ N	120 E										
2. 15 S	75 W										
3. 65 N	25 E										
4. 8 N	82 W										
5. 20 N	100 E										
6. 10 N	20 E										
7. 35 S	55 W										
8. 0	20 E										
9. 45 N	10 E										
10. 30 S	30 E										

Exercise #2
Locating Nations Using Coordinants

Directions: Place an (x) in the box that connects each nation with its coordinants

Definition – coordinants: one of a set of numbers that determines the location of a point in the space of a given dimension

NATIONS

Coordinates Lat.	Long.	Paraguay	Colombia	Mongolia	Bangladesh	Iran	Saudi Arabia	Kenya	Zimbabwe	Mali	Germany
1. 24 N	90 E				X						
2. 60 N	100 E			X							
3. 30 N	60 E					X					
4. 50 N	10 E										X
5. 20 S	60 W	X									
6. 0	70 W		X								
7. 20 N	50 E						X				
8. 0	40 E							X			
9. 20 N	0									X	
10. 10 S	30 E								X		

52

Exercise #2
Locating Nations Using Coordinants

Directions: Place an (x) in the box that connects each nation with its coordinants

Definition — coordinants: one of a set of numbers that determines the location of a point in the space of a given dimension

NATIONS

COORDINATES Lat.	Long.	Paraguay	Colombia	Mongolia	Bangladesh	Iran	Saudi Arabia	Kenya	Zimbabwe	Mali	Germany
1. 24 N	90 E										
2. 60 N	100 E										
3. 30 N	60 E										
4. 50 N	10 E										
5. 20 S	60 W										
6. 0	70 W										
7. 20 N	50 E										
8. 0	40 E										
9. 20 N	0										
10. 10 S	30 E										

Exercise #3
Locating Nations Using Coordinants

Directions: Place an (x) in the box that connects each nation with its coordinants

Definition – coordinants: one of a set of numbers that determines the location of a point in the space of a given dimension

NATIONS

COORDINATES Lat.	Long.	Egypt	Angola	Pakistan	Russia	Turkey	New Zealand	Spain	Ecuador	U.S.A.	Cuba
1. 40 S	175 E						X				
2. 30 N	30 E	X									
3. 23 N	80 W										X
4. 45 N	115 W									X	
5. 10 S	20 E		X								
6. 40 N	0							X			
7. 30 N	70 E			X							
8. 0	80 W								X		
9. 40 N	40 E				X						
10. 50 N	110 E				X						

54

EXERCISE #3
LOCATING NATIONS USING COORDINANTS

Directions: Place an (x) in the box that connects each nation with its coordinants

Definition – coordinants: one of a set of numbers that determines the location of a point in the space of a given dimension

NATIONS

COORDINATES Lat.	Long.	Egypt	Angola	Pakistan	Russia	Turkey	New Zealand	Spain	Ecuador	U.S.A.	Cuba
1. 40 S	175 E										
2. 30 N	30 E										
3. 23 N	80 W										
4. 45 N	115 W										
5. 10 S	20 E										
6. 40 N	0										
7. 30 N	70 E										
8. 0	80 W										
9. 40 N	40 E										
10. 50 N	110 E										

Exercise #4
Locating Nations Using Coordinants

Directions: Place an (x) in the box that connects each nation with its coordinants

Definition – coordinants: one of a set of numbers that determines the location of a point in the space of a given dimension

NATIONS

COORDINATES Lat.	Long.	Indonesia	India	Australia	Japan	Canada	Mexico	Brazil	Argentina	Norway	Algeria
1. 0	110 E	X									
2. 20 N	80 E		X								
3. 10 S	50 W							X			
4. 50 S	70 W								X		
5. 30 N	0										X
6. 60 N	10 E									X	
7. 30 S	150 E			X							
8. 40 N	140 E				X						
9. 30 N	100 W						X				
10. 60 N	130 W					X					

Exercise #4
Locating Nations Using Coordinants

Directions: Place an (x) in the box that connects each nation with its coordinants

Definition – coordinants: one of a set of numbers that determines the location of a point in the space of a given dimension

NATIONS

COORDINATES Lat.	Long.	Indonesia	India	Australia	Japan	Canada	Mexico	Brazil	Argentina	Norway	Algeria
1. 0	110 E										
2. 20 N	80 E										
3. 10 S	50 W										
4. 50 S	70 W										
5. 30 N	0										
6. 60 N	10 E										
7. 30 S	150 E										
8. 40 N	140 E										
9. 30 N	100 W										
10. 60 N	130 W										

HOW TO DETERMINE DIRECTIONS ON A MAP OR GLOBE

CONTENT OBJECTIVES
Students will learn to determine directions on a map or globe by use of compass points. Students will also develop skill in the use of an atlas. Knowledge of the locations of nations and cities in the United States and in the world will be expanded.

LANGUAGE OBJECTIVES
Students will gain some familiarity with foreign terms on maps and globes.

RELATED SKILLS
Following director

MATERIALS
Lesson Outline, pen

TIME
Two 45 minute period

PARTICIPATION
Students may either work with partners or individually

PREPARATION
The teacher should make sure that the students will have atlases or maps of the United States and the world available.

Answer key to practice exercise:

1. (NW)	6. (E)
2. (SE)	7. (NE)
3. (N)	8. (SW)
4. (S)	9. (N)
5. (W)	10. (W)

Answer key to class exercise:

1. (NE)	11. (SW)
2. (SE)	12. (SE)
3. (NE)	13. (NW)
4. (NW)	14. (W)
5. (SW)	15. (E)
6. (S)	16. (NE)
7. (W)	17. (NW)
8. (SE)	18. (S)
9. (E)	19. (N)
10. (N)	20. (NW)

How To Determine Directions On A Map Or Globe

I. Practice Exercise

Determining directions from and to select sets of American cities.

Directions: Use an atlas and a compass rose as well as the key given below to find the directions from and to the following sets of cities and record your decisions.

Key: N = North, S = South, E= East, W = West, NE = Northeast, SE = Southeast, SW = Southeast, NW = Northwest

_____ 1.   Tucson, Arizona to San Francisco, California

_____ 2.   Denver, Colorado to Houston, Texas

_____ 3.   New Orleans, Louisiana to St. Louis, Missouri

_____ 4.   Atlanta, Georgia to Tallahassee, Florida

_____ 5.   Spokane, Washington to Seattle, Washington

_____ 6.   Chicago, Illinois to Cleveland, Ohio

_____ 7.   Pittsburgh, Pennsylvania to Albany, New York

_____ 8.   Detroit, Michigan to Tulsa, Oklahoma

_____ 9.   Wilmington, North Carolina to Rochester, New York

_____10.   Albuquerque, New Mexico to Flagstaff, Arizona

II. Class Exercise

Determining directions from and to selected sets of world cities.

Directions: Use an atlas and a compass rose as well as the key shown in I. PRACTICE EXERCISE of this lesson to find the directions from and to the following sets of world cities.

_____ 1.   Buenos Aires, Argentina to Rio de Janeiro, Brazil

_____ 2.   Asuncion, Paraguay to Porto Alegre, Brazil

_____ 3.   Quito, Ecuador to Caracas, Venezuela

_____ 4.   Mexico City, Mexico to Mazatlan, Mexico

_____ 5.   Montreal, Canada to Toronto, Canada

_____ 6.   Algiers, Algeria to Lagos, Nigeria

_____ 7.   Newcastle, Australia to Perth, Australia

_____ 8.   Yangon, Myanmar to Bangkok, Thailand

_____ 9.   Istanbul, Turkey to Baku, Azerbaijan

_____10.   Odessa, the Ukraine to St. Petersburg, Russia

_____11.   Zurich, Switzerland to Lisbon, Portugal

___12.   Rome, Italy to Athens, Greece

___13.   Paris, France to Dublin, Ireland

___14.   Vienna, Austria to Munich, Germany

___15.   Birmingham, Great Britain to Warsaw, Poland

___16.   Copenhagen, Denmark to Helsinki, Finland

___17.   Budapest, Hungary to to Prague, the Czech Republic

___18.   Bogota, Colombia to San Jose, Costa Rica

___19.   La Paz, Bolivia to Caracas, Venezuela

___20.   Montevideo, Uruguay to Tucuman, Argentina

Title: FACT SHEET ON A NATION

Subject Area: geography

Skills: library research, analyzing

Completion Time: 45 minutes class time plus appropriate homework periods

Objective: For students to be exposed to a wide range of library reference materials

Comments: Students should read over the list of items before doing the assignment. Any questions on the lesson should be resolved in class before any work is done in the library. Many students don't realize that a nation's major agricultural product might not necessarily be one of its major exports.

Materials: pen, paper

References: encyclopedias, atlases, almanacs, magazine, books

Extension Activities: Students could classify the 32 items on the Fact Sheet on a Nation under headings like Geography, Economics, Government, Sociology, by listing the appropriate numbers of the items under the subject headings.

Notes:

FACT SHEET ON A NATION

Directions: Use library reference materials in your attempt to complete this Fact Sheet on a Nation. Some facts may not be possible to find in your library's references but hunt down as much of the required information as you can.

1. Name of nation:_____

2. Area in square miles:_____

3. Population: _____ Year reported: _____

4. Percentage of the population that is urban: _____

5. Population density per square mile: _____

6. Ethnic groups (include percentages of the population):

_____   _____

_____   _____

7. Official language(s): _____

8. Major religions (include percentages of the population): _____
_____
_____

9. Monetary unit: _____

10. Life expectancy: _____

11. Infant mortality rate: _____

12. Number of doctors per 1000 people: _____

13. Literary rate: _____

14. Number of TV sets: _____

15. Number of radios: _____

16. Number of automobiles: _____

17. Number of telephones: _____

18. Leading industries: _____

_____

_____

19. Major imports: _____

20. Major exports: _____

21. Major imports: _____

22. Major agricultural products: _____

23. Major mineral resources: _____

_____

24. Gross national product: _____

25. Per capita income: _____

26. Capital city: _____

27. Other major cities: _____

_____

_____

28. Form of government: _____

29. Bordering neighborhoods: _____

_____

30. Dominant climate type(s): _____

_____

31. Major rivers: _____

32. Membership in international organizations: _____

_____

Title: AREA STUDY QUESTION LIST

Subject Area: geography

Skills: analyzing, library research

Completion Time: 90 minutes

Objective: For students to be able to grasp the concept of an area as a subdivision of earth's surface in geographic studies

Comments: Students can make suggested additions, deletions, and/or modifications to the given list for this lesson before they undertake the assignment.

Materials: pen, paper

References: encyclopedias, almanacs

Extension Activities: Area Study Question Lists could be displayed at the end of the term. Students could make up their own examinations and provide answers so that the teacher could put together a master test for the whole class.

Notes:

Area Study Question List

Definition: An area is any tract of the earth's surface with characteristics, either natural or man made, which make it different from the areas that surround it.

Directions: Using library reference materials:
1. Define the are you are studying in terms of the nation or nations found there.
2. Write a narrative description of the area.
3. On an outline map of the world, indicate the location of the area
4. Answer the question in the Key Question List

Key Question List

A. Geographical Question on _____

    1. area in sq. mi. and kilometers _____/_____

    2. population _____

    3. population destiny (sq. mi.) _____

    4. population distribution _____

    _____

    5. population growth rate _____

    6. topography_____

    _____

    7. fauna (5 examples) _____

    _____

    8. flora (5 examples) _____

    _____

    9. neighboring areas _____

    10. chief cities_____

    _____

    11. major land forms (mountains, valleys, etc.) _____

    _____

    _____

    12. major water forms (rivers, lakes, etc.) _____

    _____

B. Historical and Political Questions on _____

    1. historical beginnings _____

    _____

    _____

    2. major events in the history of the area_____

    _____

    _____

    _____

    3. leading personalities of the area: <u>Past:</u> _____

    _____

    <u>present:</u> _____

    _____

    4. present form of government_____

    5. current events_____

    _____

    _____

C. Economic Questions on_____

    1. GNP_____

    2. major crops _____

    3. chief resources _____

    _____

    4. major industries _____

    _____

    5. chief imports _____

    6. chief exports_____

    7. per capita income_____

D. Cultural Questions on_____

    1. religions _____

    _____

    2. languages_____

    _____

    3. artistic expressions (music, painting, drama, etc.)_____

    _____

Title: COMPASS ROSE EXERCISE

Subject Area: geography

Skills: drafting, following directions, direction orientation

Completion Time: 45 minutes

Objective: For students to grasp the concept of compass directions

Comments: A compass rose is a traditional element of the map maker's craft in joining art and science in showing compass directions on a map or globe. Students should use pencil at first and then use ink when all erasing of incorrect or superfluous lines are finished.

Materials: pencil, pen, colored pencils, ruler, paper

References: atlases

Extension Activities: Enlarged compass roses could be colored and prepared for display. Students could try their hand at making innovative compass rose designs.

Notes:

COMPASS ROSE EXERCISE

A "compass rose" or a somewhat similar figure is usually found on maps to indicate directions in terms of compass points.

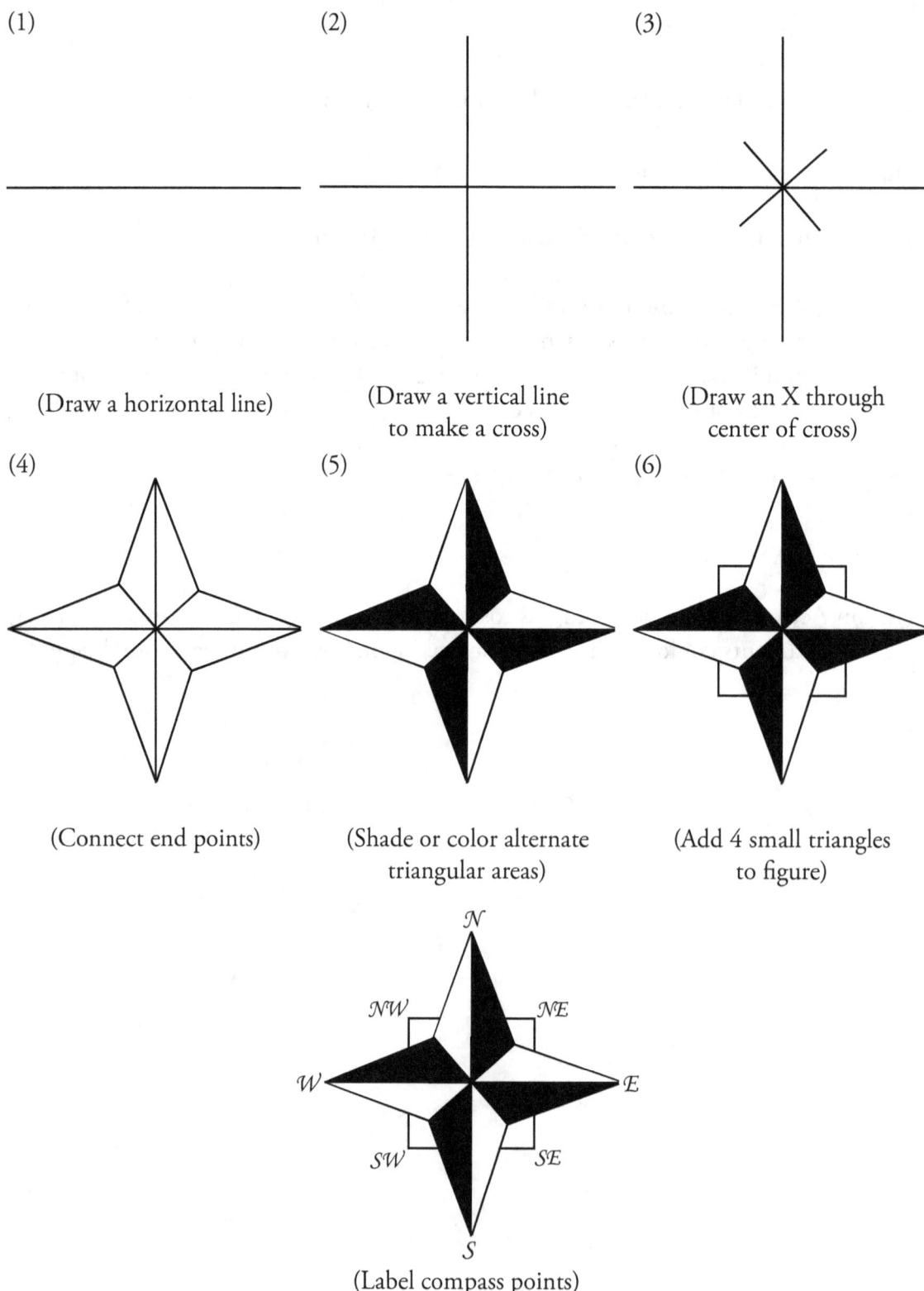

(1)

(Draw a horizontal line)

(2)

(Draw a vertical line to make a cross)

(3)

(Draw an X through center of cross)

(4)

(Connect end points)

(5)

(Shade or color alternate triangular areas)

(6)

(Add 4 small triangles to figure)

(Label compass points)

<u>Title:</u> OUTLINE FOR A REPORT ON A CITY

<u>Subject Area:</u> geography

<u>Skills:</u> library research, analyzing

<u>Completion Time:</u> 45 minutes class time plus appropriate homework periods

<u>Objective:</u> For students to identify the factors that should be examined in the study of an urban center

<u>Comments:</u> Students could "brainstorm" this topic and generate their own list of factors to examine before working on their report on a city.

<u>Materials:</u> pen, paper

<u>References:</u> encyclopedias, travel books and magazines

<u>Extension Activities:</u> Students could make a comparative chart of the major cities of a given area of the world.

<u>Notes:</u>

Outline For A Report On A City

1. Name of a city (other names, nicknames)

2. Location (longitude, latitude)

3. Elevation (distance above sea level)

4. Area (square miles/kilometers)

5. Map (show the location of the city in relation to its nation and region)

6. Current population

7. Brief physical description of the city

8. Major business, governmental, and cultural activities

9. Interesting historical information

10. Current events (within the last 10 years)

11. Tourist attractions

12. Cost of transportation to the city from your community (one way and round trip by car, train, plane, or ship)

13. Source(s) of information

Title: A PLAN FOR A CITY, STATE, REGION, NATION, OR WORLD TOUR

Subject Area: geography

Skills: library research, evaluating, synthesizing

Completion Time: 90 minutes of class time plus appropriate homework periods

Objective: For students to organize information sequentially

Comments: An example of a tour might be circulated in the class so as to provide a possible model of the lesson for the students.

Materials: pen, paper

References: travel brochures, travel books, encyclopedias

Extension Activities: Students could make an illustrated commercial for television viewing based on their selected tour assignments.

Notes:

A Plan For A City, State, Region, Nation, Or World Tour

Directions:

1. Select a city, state, region, nation or world tour that you would like to plan for a given individual or group. Clearly identify your choice in the title of your report.

2. Investigate the attractions of your selected topic so that you'll know what might be of interest to prospective tourists. List of attractions and write a brief paragraph in which you describe each attraction.

3. Determine through your research just about how long your tour might be and make a time schedule and itinerary for visiting your chosen attractions. Try to avoid extremes. You don't want the trip to drag but you don't want to exhaust your customers either.

4. Look up travel and hotel information for your trip and be sure to list all prices that you can find.

5. Establish the cost per person for your tour based on the findings from your research done using travel books and pamphlets from travel agencies.

6. List all your sources of information for your report.

Title: OUTLINE FOR A REPORT ON A PHYSICAL FEATURE

Subject Area: geography

Skills: library research, analyzing

Completion Time: 45 minutes

Objective: For students to analyze the geographical importance of a major physical feature on the earth's surface

Comments: Students might be provided with a listing of important physical features from which to make their selections for research.

Materials: pen, paper

References: encyclopedias

Extension Activities: Students could make a class display of drawings of the physical features that they have studied.

Notes:

Outline For A Report On A Physical Feature

1. Name of physical feature (a particular river, lake, mountain, etc.).

2. Description of the physical feature.

3. Map showing location of the physical feature.

4. Comparative graph showing how the selected physical feature compares with others like it in the world. e.g. The Mediterranean Sea compared with other Seas inthe world.

5. The importance of the physical feature: What events have been associated with the feature? What famous people have been connected with this feature?

6. Sources of information:

Title: OUTLINE FOR A REPORT ON A REGION

Subject Area: geography

Skills: library research, analyzing

Completion Time: 45 minutes class time plus a night's homework

Objective: For students to recognize the characteristics of a geographical region of the world

Comments: It might be a good idea to discuss some major regions of the world and to develop a list of the regions before issuing assignments to the students.

Materials: pen, paper, typewriter

References: encyclopedias, textbooks in geography

Extension Activities: A comparative chart of the world's major regions might be made by a group of students using data collected from their classmates' reports.

Notes:

Outline For A Report On A Region

Explore the characteristics of a region defined as a territorial subdivision of an established nation in the world, e.g., Katanga (Zaire), Bavaria (Germany), the Negev (Israel).

1. Name of nation and region:

2. Map of the region:

3. History of the region:

4. Economy of the region:

5. Geography of the region:

6. Social life in the region (education, health, etc.):

7. Sources of information:

Title: AN ATLAS PROJECT

Subject Area: geography

Skills: library research

Completion Time: 45 minutes of class time plus a week's homework

Objective: For students to explore a variety of map types used in geography

Comments: Students using class atlases should investigate the various kinds of maps required in the assignment before beginning their work.

Materials: pen, paper, tracing paper

References: atlases, textbooks in geography

Extension Activities: Atlases should be displayed. Excellent examples of each map type could be featured.

Notes:

An Atlas Project

Select a nation and then create an atlas of maps following the given outline.

1. Title page: Atlas of _____
    (selected nation)

2. Table of Contents:

Title: IMPRESSIONS OF A NATION

Subject Area: geography

Skills: inferring, hypothesizing, library research

Completion Time: 90 minutes

Objective: For students to recognize the need to support opinions and impressions with facts

Comments: The words at each end of the bars (continua) should be defined in class so that the possibility of student confusion might be reduced.

Materials: pen, paper

References: encyclopedias, textbooks in geography

Extension Activities: Students could write a composition entitled Impressions of _____. based on their assigned topics.

Notes:

IMPRESSIONS OF A NATION

Directions:

1. Choose a nation in the world and then look over the given bars (continua) in this lesson.

2. Notice that each bar shows two completely opposite set of factors at its end points. All intermediate factors are to be found between the extremes.

3. Use your impressions, opinions, and knowledge about the nation to place short vertical lines at points on each bar that you feel are appropriate to indicate where the nation stands in relation to the given set of factors.

4. When you have finished marking the bars, go to the library and find at least one source that either supports or refutes your marking of the bars.

5. Write a paragraph indicating whether or not your research generally supported or refuted your impressions about the nation you chose. Be sure to cite your sources.

IMPRESSIONS OF _____

Flat land		Mountains
Agricultural		Industrial
Sparsely Populated		Densely Populated
Rural		Urban
Modern		Traditional
Few Natural resources		Many natural resources
Self-sufficient		Dependent
Poverty		Prosperity
Illiteracy		Literacy
Skilled workers		Unskilled workers
Little food		Much food
Pacific		Aggressive (in war or business)
Good Public Health		Poor public health

Title: OUTLINE FOR A REPORT ON A NATION

Subject Area: geography

Skills: library research, analyzing

Completion Time: 2 weeks of homework

Objective: For students to identify the factors that should be studied in reporting on a nation

Comments: Each item in the outline should be discussed in class before students begin their work. Also, the reports should be checked for progress at the end of the first work of the assignment period.

Materials: pe, paper, typewriter

References: encyclopedias, almanacs

Extension Activities: The student reports could go into a class reference library to be used in a variety of assignments. Graphs and charts could be constructed using the students' research findings. Each student could be designated as an expert on his or her nation.

Notes:

OUTLINE FOR A REPORT ON A NATION

I. Title Page

On this page, include your name, grade, and advisor along with the name of the nation you have selected or have been assigned.

II. Table of Contents

Page

1. Physical Map of .... (Include mountains, lakes, rivers, bays, seas, etc. Show elevation, longitude, and latitude.)

2. Economic Map of .... (Show chief products of the nation along with industries in their proper locations. Indicate the chief imports and exports.

3. Political Map of .... (Show the principal cities. Star the capital. If possible, show the political divisions within the nation. Always show neighboring states.)

4. Geographic Narrative on .... (Discuss the nation's topography, its area and geographical position in the world. Also discuss its geographical advantages/disadvantages.)

5. Historical Outline of .... (Write a brief outline of major events in the nation's history and make a time line for them.)

6. Economic Outline for ...... (Discuss the nation's principal crops and industries and their importance in world trade. Go into some details about one or more crops and/or industries.)

7. Demographic Narrative on... (Give the population and its distribution, racial composition, ethnic groups, occupation, customs, foods, etc. A population map could be made for extra credit.)

8. Government of................. (Write a brief description of the nation's government and include a picture or sketch of its flag and coat of arms.)

9. Current Events in .... (Include internal and external events of importance to the nation and discuss its foreign policy.)

10. Social Narrative on .... (Discuss the nation's education, religion, arts, crafts, architecture, military organizations, sports, costumes, films, or any appropriate subject of interest to you that deals with the social life of the nation.)

11. Personalities of .... (Choose two or more notable people of this nation to write about in capsule form.)

12. Comparative Area .... (Make a simple bar graph to compare the area of the nation with those of its immediate and/or nearest neighbor.)

13. Comparative Population .... (Make a simple bar graph to compare the population of the nation with those of its immediate and/or nearest neighbor.)

14 & 15 .... (Use these two pages for illustrations of the nation.)

Title: MATCHING RIVERS WITH TRAVEL BETWEEN CITIES

Subject Area(s): geography, world history

Skills: classifying, library research, map skills

Objective(s): for students to be able to locate in the United States using a wall map as an introduction to this lesson

Materials: student sheets, pen, pencil

References: encyclopedias, almanacs, atlases, textbooks in world geography

Evaluation: less than 7 correct entries is failing

Extension Activities: students could expand or the list of rivers connecting sets of cities given in this lesson

Notes:

MATCHING RIVERS WITH TRAVEL BETWEEN CITIES

CITY TO CITY ROUTES

RIVERS	Vienna to Budapest	Khartoum to Cairo	Baghdad to Basra	Turin to Piacenza	Timbuktu to Niamey	Dawson to White Horse	Chungking to Whuan	Bonn to Rotterdam	Aliahbad to Benares	Kisangani to Brazzaville	Cordova to Sevilla	Volgograd to Astrkhan
1. Ganges									X			
2. Nile		X										
3. Yangtze							X					
4. Rhine								X				
5. Volga												X
6. Guadalquivir											X	
7. Po				X								
8. Danube	X											
9. Kongo										X		
10. Yukon						X						
11. Tigris			X									
12. Niger					X							

MATCHING RIVERS WITH TRAVEL BETWEEN CITIES

CITY TO CITY ROUTES

RIVERS	Vienna to Budapest	Khartoum to Cairo	Baghdad to Basra	Turin to Piacenza	Timbuktu to Niamey	Dawson to White Horse	Chungking to Whuan	Bonn to Rotterdam	Aliahbad to Benares	Kisangani to Brazzaville	Cordova to Sevilla	Volgograd to Astrkhan
1. Ganges												
2. Nile												
3. Yangtze												
4. Rhine												
5. Volga												
6. Guadalquivir												
7. Po												
8. Danube												
9. Kongo												
10. Yukon												
11. Tigris												
12. Niger												

Title: NATIONS AND THEIR NEIGHBORS

Subject Area(s): geography, world history

Skills: classifying, library research, map skills

Completion Time: one standard 45 minute period plus one night's homework

Objective(s): for students to recognize the geographical relationships among nations in the world

Comments: students could discuss the neighboring nations of the United States as well as the neighboring states of their state as an introduction to this lesson.

Materials: student sheets, pen, pencil

References: encyclopedias, almanacs, atlases, textbooks in world geography

Evaluation: less than 7 correct entries is failing

Extension Activities: students could expand on the list of nations and their neighbors given in this lesson

Notes:

NATIONS AND THEIR NEIGHBORS

NEIGHBORING NATIONS

NATIONS	Russia	Greece	Nepal	Angola	Panama	Argentina	Thailand	Spain	Botswana	Algeria	Romania	Honduras
1. Norway	X											
2. Nicaragua												X
3. Colombia					X							
4. Tunisia										X		
5. Bulgaria											X	
6. Albania		X										
7. Zimbabwe									X			
8. Portugal								X				
9. India			X									
10. Zaire				X								
11. Malaysia							X					
12. Paraguay						X						

Nations And Their Neighbors

Neighboring Nations

Nations	Russia	Greece	Nepal	Angola	Panama	Argentina	Thailand	Spain	Botswana	Algeria	Romania	Honduras
1. Norway												
2. Nicaragua												
3. Colombia												
4. Tunisia												
5. Bulgaria												
6. Albania												
7. Zimbabwe												
8. Portugal												
9. India												
10. Zaire												
11. Malaysia												
12. Paraguay												

Title: CITIES AND THEIR FEATURES

Subject Area(s): geography, world history

Skills: classifying, library research, map skills

Completion Time: one standard 45 minute period plus one night's homework

Objective(s): for students to recognize key features in famous cities of the world

Comments: students could use a wall map to point out the locations of the cities listed in this lesson ..... students could discuss key features in their city or nearby as an introduction to this lesson

Materials: student sheets, pencil, pen

References: encyclopedias, almanacs, atlases, books on individual cities

Evaluation: less that 7 correct entries is failing

Extension Activities: students could expand the list of cities and their features

Notes:

CITIES AND THEIR FEATURES

CITIES	Eiffel Tower	The Kremlin	The Imperial City	The Ginza	Copacabana Beach	The Zocalo	The Shwe Dagon Pagoda	Santa Sophia	The Reichstag	The Prado	Raffles Hotel	United Nations Headquarters
1. Istanbul								X				
2. Moscow		X										
3. Madrid										X		
4. Paris	X											
5. Rio de Janeiro					X							
6. Singapore											X	
7. Peking			X									
8. Berlin									X			
9. New York												X
10. Tokyo				X								
11. Mexico City						X						
12. Rangoon							X					

CITIES AND THEIR FEATURES

FEATURES OF CITIES

CITIES	Eiffel Tower	The Kremlin	The Imperial City	The Ginza	Copacabana Beach	The Zocalo	The Shwe Dagon Pagoda	Santa Sophia	The Reichstag	The Prado	Raffles Hotel	United Nations Headquarters
1. Istanbul												
2. Moscow												
3. Madrid												
4. Paris												
5. Rio de Janeiro												
6. Singapore												
7. Peking												
8. Berlin												
9. New York												
10. Tokyo												
11. Mexico City												
12. Rangoon												

Title: NATIONS AND THEIR POLITICAL SUBDIVISIONS

Subject Area(s): geography, civics, world history

Skills: classifying, library research, map skills

Completion Time: one standard 45 minute period plus one night's homework

Objective(s): for students to be able to identify important political subdivisions in the nations of the world

Comments: the students could be called upon to point out the political subdivisions listed in this lesson on a wall map

Materials: student sheets, pen, pencil

References: encyclopedias, almanacs, atlases, textbooks in geography and world history

Evaluation: less than 7 correct entries is failing

Extension Activities: students could expand the list of nations and their political subdivisions given in this lesson

Notes:

NATIONS AND THEIR POLITICAL SUBDIVISIONS

NATIONS	Bavaria	Manitoba	Chiapas	Oriente	Santa Catarina	Leitrim	Orissa	Hunan	West Kasai	Orange Free State	Queensland	Liguria
1. Australia											X	
2. Canada		X										
3. Brazil					X							
4. Mexico			X									
5. Democratic Republic of the Congo									X			
6. Italy												X
7. Cuba				X								
8. South Africa										X		
9. Germany	X											
10. China								X				
11. Ireland						X						
12. India							X					

Nations And Their Political Subdivisions

Nations	Bavaria	Manitoba	Chiapas	Oriente	Santa Catarina	Leitrim	Orissa	Hunan	West Kasai	Orange Free State	Queensland	Liguria
1. Australia												
2. Canada												
3. Brazil												
4. Mexico												
5. Democratic Republic of the Congo												
6. Italy												
7. Cuba												
8. South Africa												
9. Germany												
10. China												
11. Ireland												
12. India												

Title: NATIONS AND THEIR PHYSICAL FEATURES

Subject Area(s): geography, world history

Skills: classifying, library research, map skills

Completion Time: one standard 45 minutes period plus one night's homework

Objectives(s): for students to identify important world physical features

Comments: the teacher could go over some basic terms of physical geography with the students as an introduction to this lesson

Materials: student sheets, pen, pencil

References: encyclopedias, almanacs, atlases, textbooks in the world geography

Evaluation: less than 7 correct entries is failing

Extension Activities: students could expand the list of nations and their physical features given in this lesson

Notes:

Nations And Their Physical Features

Nations	Eastern Chats	Dashte Kavir Desert	Lake Toniesap	Mt. Apo	Great Sandy Desert	Bornu Plains	Okavango Swamp	Great Slave Lake	Isthmus of Tehuantepec	Madre de Dios Archipelago	Ustyurt Plateau	Lizard Point
1. Chile										X		
2. Iran		X										
3. United Kingdom												X
4. India	X											
5. Nigeria						X						
6. Cambodia			X									
7. Kazakhstan											X	
8. The Philippines				X								
9. Australia					X							
10. Mexico									X			
11. Botswana							X					
12. Canada								X				

Nations And Their Physical Features

Nations	Eastern Chats	Dashte Kavir Desert	Lake Toniesap	Mt. Apo	Great Sandy Desert	Bornu Plains	Okavango Swamp	Great Slave Lake	Isthmus of Tehuantepec	Madre de Dios Archipelago	Ustyurt Plateau	Lizard Point
1. Chile												
2. Iran												
3. United Kingdom												
4. India												
5. Nigeria												
6. Cambodia												
7. Kazakhstan												
8. The Philippines												
9. Australia												
10. Mexico												
11. Botswana												
12. Canada												

Title: SOURCE SEARCH FOR A REPORT ON A NATION

Subject Area: general social studies

Skills: library research

Completion Time: 45 minutes

Objective: For students to investigate the available library resources for a report on a nation in the world

Comments: Before students begin a report on a nation, the teacher should inform the librarian so that materials can be organized for the students. The librarian should be given a copy of the lesson to make things easier. This Source Search is a good way for students to start working on their reports – much student anxiety can be blunted by their knowing that materials on their selected assignment are available.

Materials: pen, paper

References: library reference materials on the world's nations

Extension Activities: The class could discuss their reactions to the variations reference materials that they studied.

Notes:

Source Search For A Report On A Nation

Name of a nation: _____

Directions: Look up the coverage of the selected nation in each of the
following books:

1. <u>World Almanac</u>: year _____, page_____ to _____

2. <u>Worldmark Encyclopedia of Nations</u>: year _____, page _____
    to _____

3. <u>Encyclopedia Americana</u>: year_____, page _____ to_____

4. <u>World Book Encyclopedia</u>: year_____, page _____ to _____

5. A Comprehensive World Atlas of your choice: year _____, page_____

6. A Book (total or partial coverage of the nation):

    Title of the book:_____

    Author_____

    Place of publication: _____

    Date of publication: _____

    Number of pages devoted to a coverage of the nation _____

Teacher's signature:_____ Date signed:_____

Date assignment is due: _____

Title: POLITICAL CHANGE IN AFRICA

Skills: translating, interpreting

Completion Time: one 45 minute class period

Objective: For students to interpret a political map

Comments: The teacher should mention how the result of World War I shifted the European domination of Africa more to France and Great Britain.

Answer Key to Map Exercise Sheet #2

1. Madagascar (French)
2. Algeria (French)
3. Libya (Italian)
4. Equatorial Guines (Spanish)
5. Angola (Portuguese)
6. Namibia (British)
7. Tanzania (German)
8. Chad (French)
9. Democratic Republic of the Congo (Belgian)
10. Nigeria (British)
11. Ghana (British)
12. Liberia (Independent)
13. Ethiopia (Independent)
14. Cameroon (German)
15. Somalia (Italian/British)
16. Botswana (British)
17. Zambia (British)
18. Mali (French)
19. Western Sahara (Spanish)
20. Chad (French)
21. Tunisia (French)
22. Zimbabwe (British)
23. Kenya (British)
24. Sudan (British)
25. Mozambique (Portuguese)

Materials: Map Exercise sheets #1 & #2, pen or pencil

References: almanacs, encyclopedias

Extension Activities: Students could create an exercise for Asian nations modeled on this one for Africa.

Notes:

MAP EXERCISE SHEET # 1

British
Spanish
French
Belgian
German
Portuguese
Italian
Independent

AFRICA IN THE LATE 19ᵀᴴ CENTURY

AFRICA TODAY

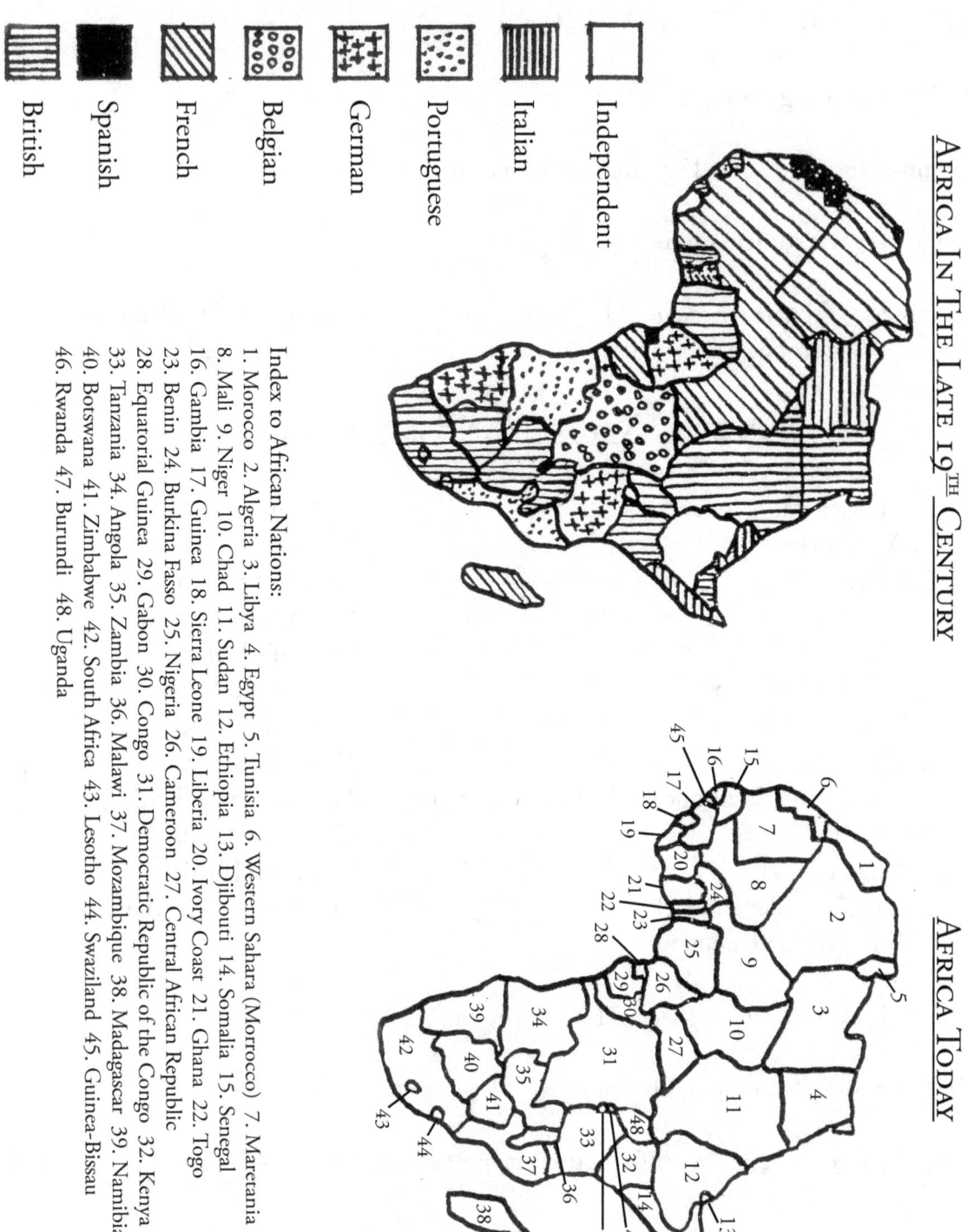

Index to African Nations:
1. Morocco 2. Algeria 3. Libya 4. Egypt 5. Tunisia 6. Western Sahara (Morocco) 7. Mauretania
8. Mali 9. Niger 10. Chad 11. Sudan 12. Ethiopia 13. Djibouti 14. Somalia 15. Senegal
16. Gambia 17. Guinea 18. Sierra Leone 19. Liberia 20. Ivory Coast 21. Ghana 22. Togo
23. Benin 24. Burkina Fasso 25. Nigeria 26. Cameroon 27. Central African Republic
28. Equatorial Guinea 29. Gabon 30. Congo 31. Democratic Republic of the Congo 32. Kenya
33. Tanzania 34. Angola 35. Zambia 36. Malawi 37. Mozambique 38. Madagascar 39. Namibia
40. Botswana 41. Zimbabwe 42. South Africa 43. Lesotho 44. Swaziland 45. Guinea-Bissau
46. Rwanda 47. Burundi 48. Uganda

MAP EXERCISE SHEET #2

Directions: Fill the chart that follows:

CHANGES OF POLITICAL STATUS IN AFRICAN NATIONS

African Nations Today	Political Status of African Nations in Late 19th Century
1. Madagascar	
2. Algeria	
3. Libya	
4. Equatorial Guinea	
5. Angola	
6. Namibia	
7. Tanzania	
8. Chad	
9. Democratic Republic of the Congo	
10. Nigeria	
11. Ghana	
12. Liberia	
13. Ethiopia	
14. Cameroon	
15. Somalia	
16. Botswana	
17. Zambia	
18. Mali	
19. Western Sahara	
20. Chad	
21. Tunisia	
22. Zimbabwe	
23. Kenya	
24. Sudan	
25. Mozambique	

Title: GRAPHIC RELATIONSHIPS: CLASSIFYING ITEMS BY USING QUADRANTS

Subject Area: geography

Skills: graphing, classifying

Completion Time: one 45 minute period in the school library

Objective: For students to graphs some geographical relationships

Comments: Students need to have the graphs explained to the along with the concept of quadrants. The teacher might give the students one of the four items in each graph to help the students come up with the other three.

Materials: pen, pencil, handouts

References: encyclopedias, almanacs

Extension Activities: Students could write a brief report on my one item from any of the quadrant in any of the graphs

Notes:

Answer Sheet for Teachers:

1. Classification of Nations: by land area and population

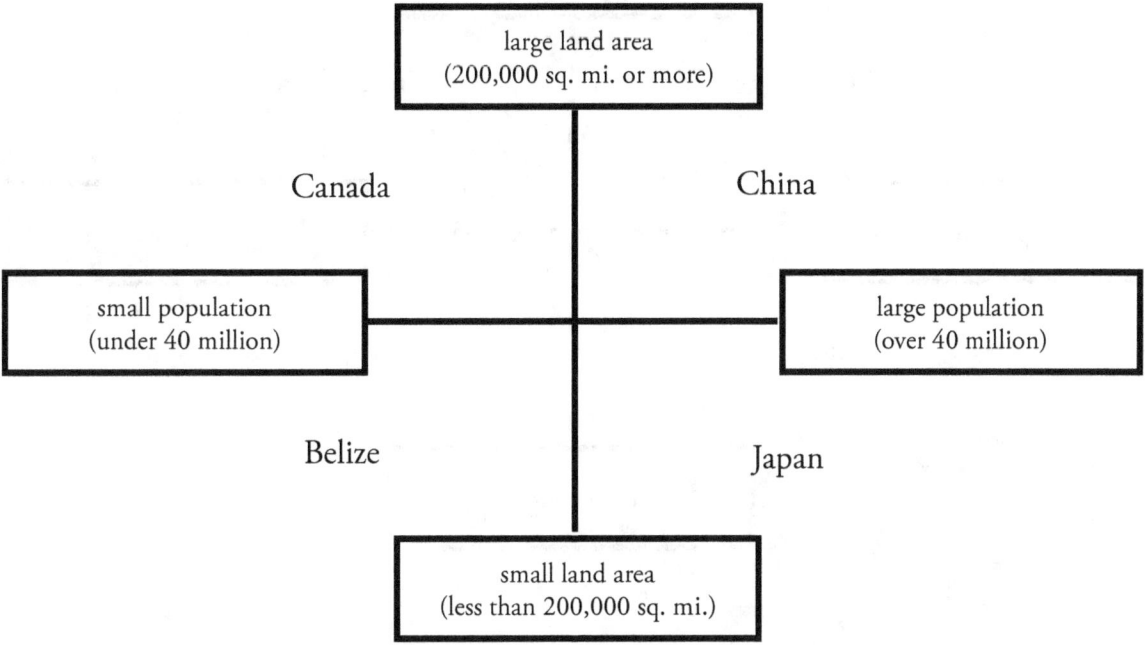

large land area
(200,000 sq. mi. or more)

Canada   China

small population   large population
(under 40 million)   (over 40 million)

Belize   Japan

small land area
(less than 200,000 sq. mi.)

2. Classification of Cultures by climatic conditions:

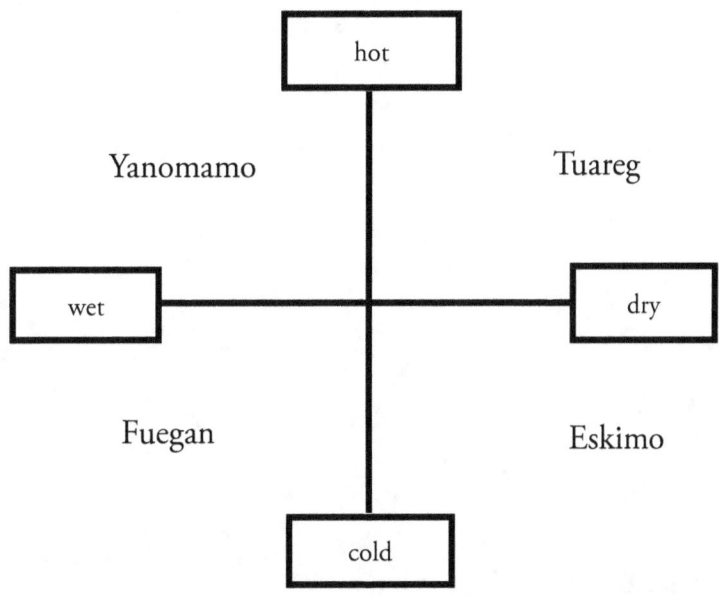

hot

Yanomamo   Tuareg

wet   dry

Fuegan   Eskimo

cold

3. <u>Classification of nations by reference to selected economic factors:</u>

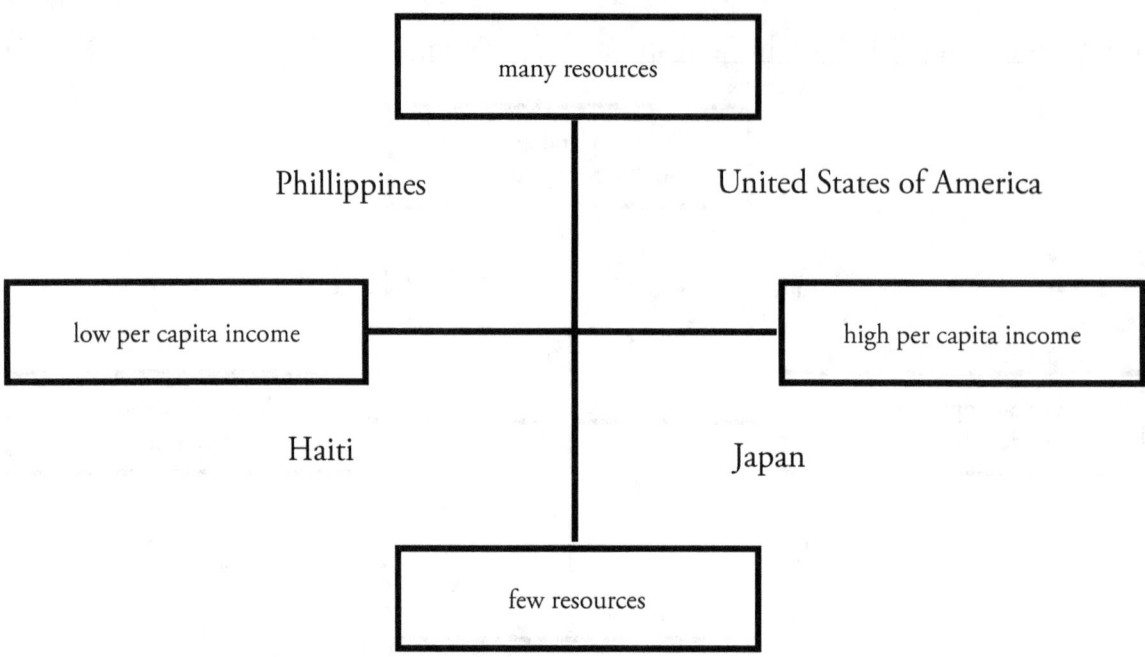

STUDENT GRAPH EXERCISE

<u>Directions:</u> list one item in each quadrant of the following graphs

1. <u>Classification of nations by land area and population:</u>

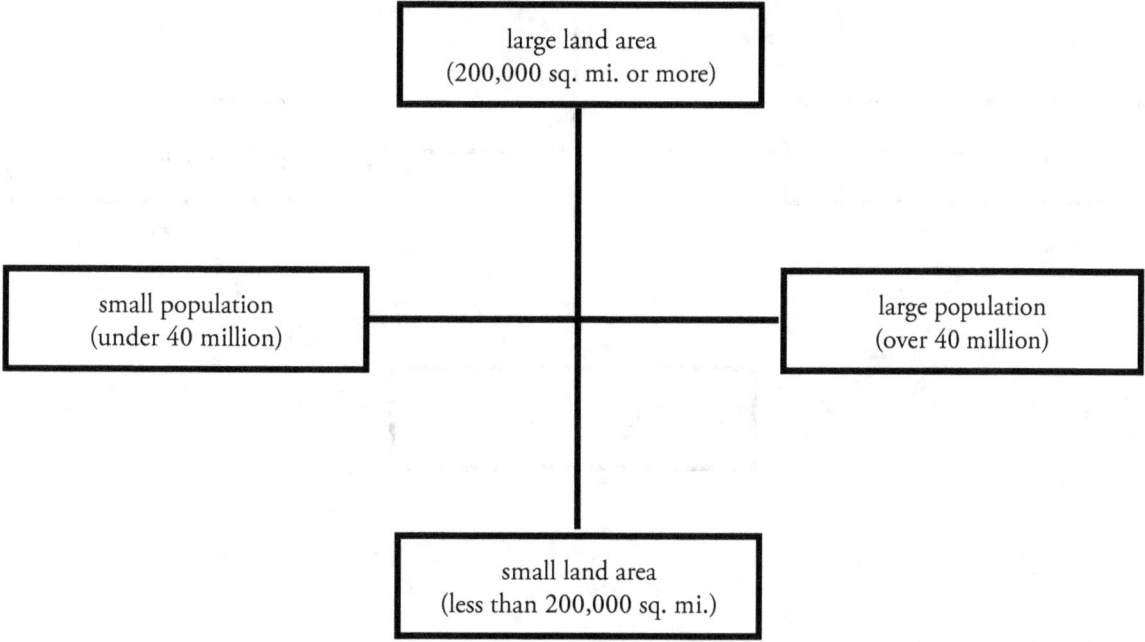

2. <u>Classification of cultures by climatic conditions:</u>

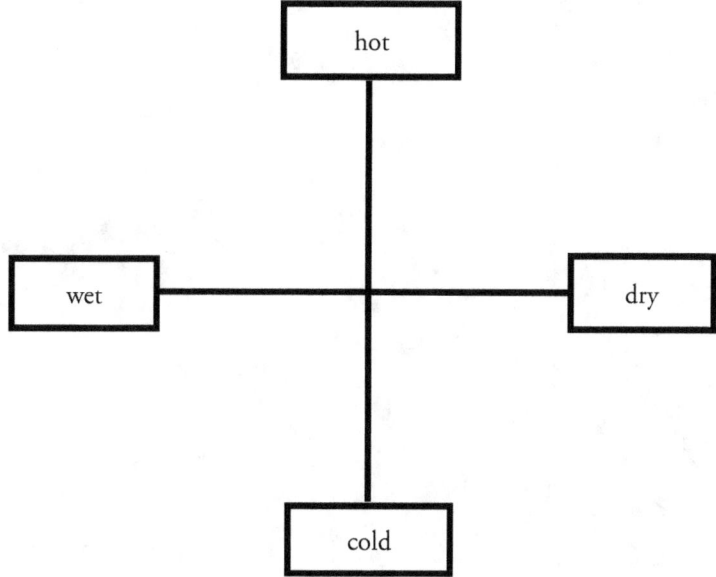

3. <u>Classification of nations by reference to selected economic factors:</u>

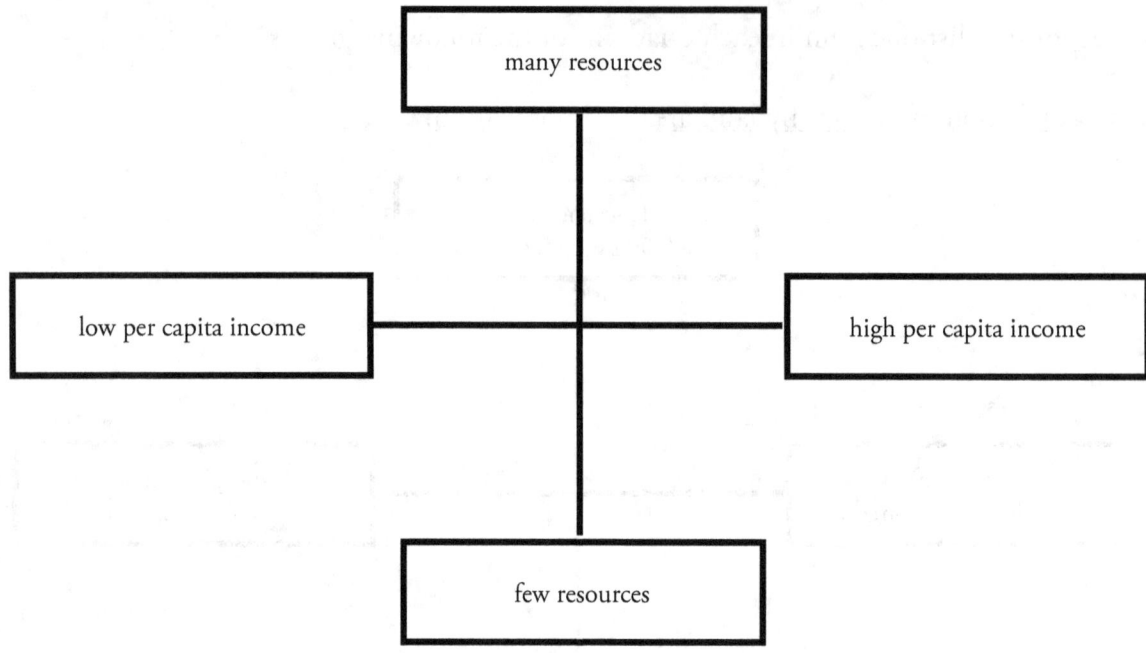

Title: USING GRID CHARTS IN GEOGRAPHY

Subject Area: geography

Skills: classifying, scanning, analyzing, spatial recognition

Completion Time: two 45 minute periods in the school library or classroom if atlases are available to the students

Objective: for students to develop skill in working with grid charts

Comments: students will use wall maps to point out hemispheres, continents, nations, regions, states and cities to their classmates as an introduction to the two exercises in this lesson

Definition of the term "grid": a pattern of horizontal and vertical lines forming squares of uniform size on a map or chart

Materials: exercise sheets, pen or pencil

References: atlases, almanacs

Extension Activities: the teacher can create a new grid chart assignment for the students to do based on the exercise in this lesson

Notes:

ANSWER KEYS FOR EXERCISES A AND B:

Exercise A answer key:

THE HEMISPHERES: AN EXERCISE IN LOCATION Directions: For each item below, place two (x's) that show the two hemispheres that describe the general location of the item	Northern Hemisphere	Southern Hemisphere	Western Hemisphere	Eastern Hemisphere
1. Canada	X		X	
2. Australia		X		X
3. Chile		X	X	
4. Japan	X			X
5. Argentina		X	X	
6. New Zealand		X		X
7. Mexico	X		X	
8. Mozambique		X		X
9. China	X			X
10. Indonesia		X		X
11. New York City	X		X	
12. Asia	X			X
13. Angola		X		X
14. Montevideo		X	X	
15. Panama	X		X	
16. Europe	X			X
17. Haiti	X		X	
18. Cuba	X		X	
19. The Phillippines	X			X
20. Paraguay		X	X	

Exercise B answer key:

IDENTIFICATION OF GEOGRAPHICAL LAND DIVISIONS Directions: For each one of the items below, place an (x) in the box that showsis classification	Continent	Region	Nation	State/Province	City
1. Uruguay			X		
2. Southeast Asia		X			
3. Tokyo					X
4. Hawaii				X	
5. Shanghai					X
6. Antarctica	X				
7. Africa	X				
8. Scandinavia		X			
9. Manitoba				X	
10. Georgia			X	X	
11. Lagos					X
12. South America	X				
13. Thailand			X		
14. Latin America		X			
15. Katanga				X	
16. Bavaria				X	
17. Iran			X		
18. Hunan				X	
19. Calcutta					X
20. Pakistan			X		
21. Hanoi					X
22. The Phillippines			X		
23. Europe	X				
24. Oslo					X
25. Tunisia			X		

STUDENT SHEET:
TERMS FOR STUDENTS TO REVIEW:

Eastern Hemisphere = the part of the world including the continents of Europe, Africa, Asia, and Austria

Western Hemisphere = the half of the earth that includes all of North and South America the surrounding waters and all neighboring islands

Northern Hemisphere = the half of the earth north of the equator

Southern Hemisphere = the half of the earth south of the equator

continent = one of the principal land masses of the earth, usually regarded as including Africa, Antarctica, Asia, Australia, Europe, North America, and South America

region = a specified district or territory that is a somewhat indefinite portion of the earth's surface

nation = an aggregation of people organized under a single government

state = one of the more or less internall autonomous territorial and political units compromising a federation under a sovereign government

province = a territory governed as an administrative or political country or empire

city = a town of significant size

Exercise A

THE HEMISPHERES: AN EXERCISE IN LOCATION Directions: For each item below, place two (x's) that show the two hemispheres that describe the general location of the item	Northern Hemisphere	Southern Hemisphere	Western Hemisphere	Eastern Hemisphere
1. Canada				
2. Australia				
3. Chile				
4. Japan				
5. Argentina				
6. New Zealand				
7. Mexico				
8. Mozambique				
9. China				
10. Indonesia				
11. New York City				
12. Asia				
13. Angola				
14. Montevideo				
15. Panama				
16. Europe				
17. Haiti				
18. Cuba				
19. The Phillippines				
20. Paraguay				

Exercise B

IDENTIFICATION OF GEOGRAPHICAL LAND DIVISIONS Directions: For each one of the items below, place an (x) in the box that showsis classification	Continent	Region	Nation	State/Province	City
1. Uruguay					
2. Southeast Asia					
3. Tokyo					
4. Hawaii					
5. Shanghai					
6. Antarctica					
7. Africa					
8. Scandinavia					
9. Manitoba					
10. Georgia					
11. Lagos					
12. South America					
13. Thailand					
14. Latin America					
15. Katanga					
16. Bavaria					
17. Iran					
18. Hunan					
19. Calcutta					
20. Pakistan					
21. Hanoi					
22. The Phillippines					
23. Europe					
24. Oslo					
25. Tunisia					

www.ingramcontent.com/pod-product-compliance
Lightning Source LLC
Chambersburg PA
CBHW081132170526
45165CB00008B/2643

* 9 7 8 1 4 3 4 3 6 8 1 3 3 *